GW01279215

By the same author:
Victorian Theatricals, from menageries to melodrama
(Methuen)

Islomania

Sara Hudston

First published in 2000
by AGRE BOOKS
Groom's Cottage, Nettlecombe,
Bridport, Dorset, DT6 3SS

www.agrebooks.co.uk

All rights reserved. No part of this book may be reproduced in any form or by any means without permission in writing from the publisher, except by a reviewer who may quote brief passages in a review.

Copyright © 2000 Sara Hudston

The author has asserted her moral right

Typeset and printed in the British Isles
by Bigwood & Staple
Bridgwater, Somerset

ISBN 0 9538000 0 8

A CIP catalogue record for this book
is available from the British Library

Frontispiece: Islanders survey the hulk of the *James Armstrong* on the Town Beach, St Mary's in the late 1870s.
Picture by John Gibson.

For Lis and Peter, who encouraged me

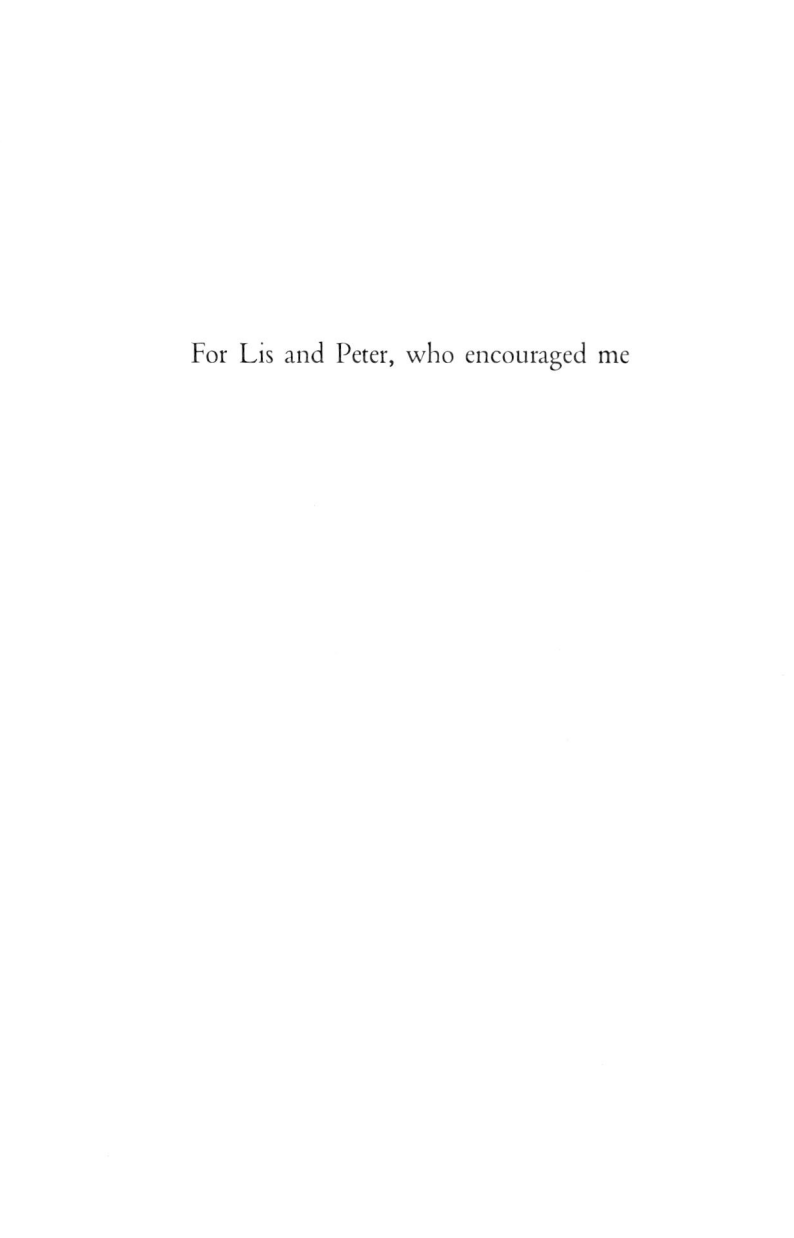

Acknowledgements

My thanks are due to the following without whom this book would not exist:

Frank Gibson for kindly allowing use of photographs from the Gibson Archive, Matt Lethbridge for spending time talking to me about his life and showing me his paintings, and his wife Pat for letting me quiz her, Clive Mumford for discussions in the Bishop and Wolf on St Mary's, Tracey Clunies-Ross for letting me interrogate her about her family history, Kate Summerscale, South West Arts for funding assistance, Stuart Brill and Senate for being so helpful and speedy with cover design, Guy Crowden for permission to reproduce the map from Camden's *Britannia*, James Crowden for cider and support, and Jonathan Hudston, who first took me to Scilly.

Extract from *The Queen of Whale Cay* reprinted by permission of Fourth Estate Ltd. Copyright © 1997 Kate Summerscale.

Look, Stranger by W.H. Auden reprinted by permission of Faber & Faber Ltd.

Gibson Archive photographs copyright © 2000 Frank Gibson.

CONTENTS

ISLANDS	5
THE SEA AND THE MIRROR	6
CROSSING THE WATER	9
WESTWARDS	13
A BIT OFF THE MAP	17
ARCHIPELAGOS OF SELF	23
LOOK, STRANGER	27
ISOLATION	28
I WANT TO GET AWAY	33
BEING AN ISLANDER	36
IN VIEW OF SCILLY	41
SO THAT STRANGERS MIGHT SEE	45
THE LEAPING LIGHT	49
ECONOMIES OF SCALE	50
THE MASTER	52
HOME RULE	61
THE ISLANDS COMPARED TO A FEAST	70
FOOLS IN NEW-STYLE HATS AND COATS	77
WHAT DO THESE PEOPLE WANT WITH THE ISLANDS?	80
CASTAWAY	84
ALONE ON A WIDE WIDE SEA	85
FURTHER READING	87

Looking towards St Mary's and St Agnes from Tresco with Cromwell's Castle in the foreground. The castle was built by Parliamentary troops in 1651. The view from the parapet over the shimmering islands is one of the best on Scilly. Picture by Frank Gibson.

LOOK, STRANGER

Look, stranger, at this island now
The leaping light for your delight discovers,
Stand stable here
And silent be,
That through the channels of the ear
May wander like a river
The swaying sound of the sea.

Here at the small field's ending pause
Where the chalk wall falls to the foam, and its tall ledges
Oppose the pluck
And knock of the tide,
And the shingle scrambles after the suck-
ing surf, and the gull lodges
A moment on its sheer side.

Far off like floating seeds the ships
Diverge on urgent voluntary errands;
And the full view
Indeed may enter
And move in memory as now these clouds do,
That pass the harbour mirror
And all the summer through the water saunter.

W.H. Auden

*Rough sea around Tooth Rock, Peninnis Head, St Mary's.
Picture by Frank Gibson.*

Islands

Be careful when you go to the Isles of Scilly – you may become an islomane.

Islomane – someone with a mania for islands. Can also mean someone sent crazy by islands, an island maniac.

Islomania is an odd word. Lawrence Durrell used it in *Reflections on a Marine Venus*, his 1953 book about Rhodes, a book he described as "an anatomy of islomania". Durrell was a confirmed islomane who spent his entire life trying to "get away". He fled from England, which he traduced as Pudding Island, from Englishness and niceness, sobriety and restraint.

Lawrence's younger, cuddlier, brother Gerald also wrote an anatomy of islomania. *My Family and Other Animals* was a lyrical description of Gerald's childhood on Corfu in the 1930s. It was a bestseller which has remained in print ever since it was published in 1956. Together the Durrell brothers more or less invented the modern idea of the island as paradise retreat, a place of sunlight shimmering off water where days "drop as softly as fruit from trees", a sacred place and time sold nowadays to us all as "holiday". Time off, time out in a separate and wholly enticing reality.

Their islands were warm and un-English in their landscape and customs. They were never lands. "Neverland is always more or less an island," wrote J.M. Barrie in *Peter Pan*.

The sea and the mirror

Of course Lawrence and Gerald destroyed what they were trying to reach. They brought package tours to the simple places they loved. More than this they spoiled things for themselves by creating impossibly lovely islands of the imagination that made travesties of the real thing. This was bound to happen, for this is the paradox of all islands; get there and they are islands no longer. Your journey has made a bridge which strings them to the mainland. We can't get to islands because they are what we are not. "No man is an *Iland*, intire of it selfe; every man is a peece of the *Continent*, a parte of the maine." John Donne was right to lecture; we are hopelessly mainland but we all fantasise about being islands.

The Durrells dreamed about finding a secret way that could take you into an island's inner world, into Neverland. Gerald thought he'd performed this mysterious Alice in Wonderland translation. He called it "passing through the mirror" and it happened in retrospect and always by sea. From his sickbed in a drab attic in Bournemouth he looked back twenty years and wrote of waking at dawn, aged ten, on a Greek ferry and finding himself inside the bright mirror of Corfu. At the same time Lawrence, embroiled in marital difficulties and exiled to communist Belgrade, imagined himself undergoing a storm on the way to Rhodes and emerging sea-tested inside a similar, shining, perfect, paradise.

Part of the Isles of Scilly: Bryher is in the foreground, then Tresco, Round Island, St Helen's, Tean and finally St Martin's. Frank Gibson took this picture in the 1960s, way before the hotel was built on Bryher in 1980.

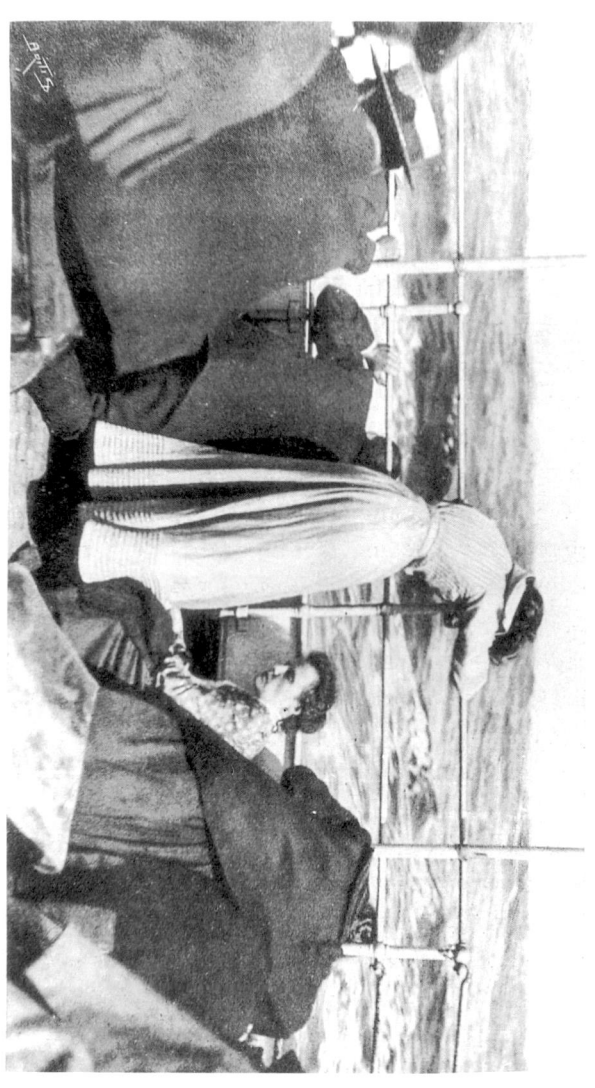

"Far, Far From Land": The anguish of seasickness. One of a series of postcards called "A Holiday Trip" taken by Alexander Gibson in the early 1900s.

Crossing the water

The Durrells were right to insist on travel by sea. For the magic to work you must approach an island on its own terms, by water. That way you learn the difference between sea and land.

It is possible to fly to Scilly on a deafening helicopter or insect-bodied plane from Penzance or Land's End. The journey takes less than twenty-five minutes and the islands whizz up from below like tilting splats of green phlegm. Everything is dinky and insignificant, you see the whole at once and land feeling disorientated. No, the way to go is by boat on the lumbering *Scillonian III* which leaves Penzance at 9.15 in the morning and takes about three hours to "cruise" thirty-five miles. Those inverted commas are a warning, the voyage is sold to day-tripping pensioners as a cruise, but it's not. True, an informative commentary is broadcast as you sail past Newlyn, Mousehole and the Minack Theatre nestling in the cliffs above Porthcurno. Then at Wolf Rock the boat starts to lift on the Atlantic rollers and the mainland drops away like so much ballast thrown overboard. The boat judders and rolls in three directions at once and the passengers stagger and clutch. For more than an hour it plunges onwards. A flying squall of rain hurries overhead to Cornwall. On deck everyone is cold and nauseous, below deck old ladies sit in the saloon with their heads in their hands, propping their elbows on the Formica tables and groaning. The low throb of the engines shudders in your spine. Dogs are unaffected. At first they peer excitedly between the deck railings at the sea rushing by, later they find a comfy place, preferably on a coat or a seat, and fall asleep.

Suddenly, you see the broken dinosaur skeletons of the Eastern Isles poking from the sea. At high tide the boat steadies and glides into the lagoon smoothly as butter on a hot dinnerplate. It eases cautiously over the sand ridge of Crow Bar and docks at St Mary's harbour. At low tide it chugs choppily between St Mary's and St Agnes, a route which has seen many disastrous wrecks. You pass Porth Hellick where in 1707 Admiral Sir Cloudesley Shovel's body was washed up after he lost his bearings in fog and struck his fleet on the treacherous Gilstone Rock. The sea still exacts its toll; further on is Peninnis Head. Here a young Cornish fisherman drowned in 1999 when his boat the *Rachel Harvey* went down two hundred yards off land.

Here's a tip to remember – when the ferry lands, totter into the pub at the end of the quay and look at the early Edwardian postcards by the bar. These old photos from the Gibson Archive show trippers making the crossing a century ago on the steamer. There's one captioned "Far, Far From Land" showing tight-waisted ladies hanging on to the rails faint with seasickness. Raise a glass to them. Out at sea you and they underwent the same experience and the century between you crumpled like starched cloth in salt spray.

The journey by sea used to take a lot longer. Before the steamship the *Little Western* was introduced in 1859 the passage by sailing boat in the *Ariadne* took twelve hours on a good crossing. In bad weather it could take as long as two days. No wonder that before the advent of steam power tourists were largely unknown on the islands. George Eliot's lover G.H. Lewes, who visited Scilly with Eliot in 1857, commented: "Vessels in abundance touch there; but who *goes* there?" Sixty years later the crossing could take up to five hours. In 1910 Jessie Mothersole described how some day-trippers arrived "in such a woebegone condition that they are fit for nothing but to lie about on the benches in the Park until the hoot of the steamer rouses them to crawl back to the quay." Until the aeroplane service began in 1937 you had to suffer to reach Scilly.

Steam Communication to Scilly.

THE STEAMER
"LADY OF THE ISLES,"
Runs regularly between Penzance and Scilly.

Average passage 4 hours.

For particulars of Sailing apply to

JOHN BANFIELD, Jun.,

Manager.

WEST CORNWALL STEAMSHIP Co.,

Offices: 6, North Parade, PENZANCE.

"Life Becomes A Burden": Spewing overboard on the steamer. Titles of the other four postcards in this series were "Trouble Brewing", "A First Taste of the Atlantic", "Effect of a Cross Sea" and finally, "A General Collapse".

Westwards

Which do you prefer, North or South? Rain and wind, occasional sunlight that ruffles the water into shivers, or sustained full-bodied sunshine, heady as a draft of Greekish wine swigged direct from a bowl?

Scilly is neither but it can pretend to be both. Scilly is West.

West is a land all to itself. It is Lyonnesse, the lost land you can chase through fairytales. To Thomas Hardy pining for his dead wife, West was a definite land of its own, a place which belonged to lost love, a love that his poems experience as a lost place. In mythology, West is always the land of the dead.

In most Britons' minds West is holidayland. Its borders are natural features; some geographic, some living. There are the moorland bulks of Exmoor, Dartmoor and Bodmin, the Quantock, Mendip and Purbeck Hills, the Tamar and the muddy surge of the Bristol Channel, punctuated with the minute table-top islands of Flat Holm and Steep Holm. The finest of the living borders is the buzzard line that divides Britain into west and east, wild and tame. There are few buzzards east of Salisbury Plain, how could there be in the flat, farmed acres of Hampshire? In the south buzzards start with Dorset, a county whose rich name reminds one of "sunset". West of the Isle of Portland, towards Lyonnesse and the setting sun, the local stone turns the golden-tawny colour of the best Dabinett cider.

Portland is the true east/west marker. Portland is the restraining foot that stops the eastward sweep of Lyme Bay and boots it back west. Geographically it is distinct, a four-and-a-half-mile wedge of marble jutting out into the English Channel. The isle has been inhabited since Neolithic times when people came to chip the fine stone into tools. It is a harsh, treeless place of rock, a natural fortress which houses two prisons. Cuckoo pint grows there in profusion and in Tudor times the plant was harvested and the dried roots used to starch lace collars. In the eighteenth and nineteenth centuries the stone was quarried to build London's landmarks – St Paul's Cathedral rose entire from a Portland quarry. The island had its own breed of web-footed dog, now extinct, and a variety of blue-black potato. Its dirty-brown sheep make excellent eating and were prized for their savour.

Portland people were notoriously insular. They hated mainlanders, calling them "kimberlins", an ancient dialect word for unwelcome strangers, and threw stones at those who tried to cross the tidal causeway. In the past, many inhabitants never left the island and it was common for them to spend entire lives in the same village. Some didn't even travel to the other end of their own small kingdom. Suspicious of strangers and other places, their worlds contracted to a few well-known things, and they constructed a mental island that was even more remote and hard to reach than the geographical one under their feet. This is one of the features of islomania.

Leonard and Fred Jenkins load a cow at Bryher to take across to Tresco. They are using a small boat because it had low sides. It was quite normal to transport livestock this way and yes, the cow did get in.
Picture by Frank Gibson, late 1940s.

Portland today is linked to the mainland by a causeway and a bridge so it's no longer really an island. Dorset has another isle which is not – the Isle of Purbeck. Like Scilly, Purbeck is confusingly plural and singular. When people speak of "Purbeck" they mean the isle, when they say "the Purbecks" they mean the Purbeck Hills. Both are correct. Those who think that geographical features alone determine a place's character had better keep it plural and call them the Purbecks, for there is indeed a range of hills swooping down to spectacular cliffs. But if you believe that the way people think about a place can shape its nature then call it an isle for it has all the distinguishing features.

Purbeck has sea on three sides and on the fourth it used to have a heath which was as treacherous and untrustworthy as the sea. This is the heath Hardy dramatised in *The Return of the Native*, a brooding wasteland of adder-sheltering furze which severed the damp, green hills of Purbeck from the outside world. Today the heath is gone and Purbeck is an easy-ish to visit holiday destination for those who can endure the traffic jams. In August upwards of thirty thousand cars a day pass through Corfe Castle but how many of their passengers really go there as they watch the landscape pass by as remote as television? Purbeck is still difficult to get to and stubbornly itself.

A bit off the map

Scilly is as west as is gets in this country. Twenty-nine miles off Land's End, or is it twenty-eight, or thirty, no one seems quite sure. The confusion is typical. Scilly is small and scattered, it is not one tidy block which can be straightforwardly charted. This makes it lethal to shipping. There are about one hundred and fifty islands, islets, rocks and ledges depending on the tide and the weather. Five islands are still inhabited, nine used to be. Once there were just two islands, and before that, before the end of the last ice age when the sea welled up and made islanders of us all, then Scilly was an outcrop of high hills on a vast plain of bison grass. Now perhaps there are fewer than one hundred and fifty islets as the sea continues to rise. On the deserted island of Tean the waves already lap the lintels of a ruined seventeenth century farmhouse and submerge the path, which used to run over rabbit-nibbled turf down to the beach. Global warming will destroy Scilly in a few hundred years if sea levels rise as high as some scientists predict.

Here on Scilly it is normal for the islands to appear and disappear. On rare occasions when the tide goes out far enough it is possible to walk from one island to another across the sandy flats. The literary critic Geoffrey Grigson did it once in the early 1940s, growing increasingly irascible as the gluey ground weighed down his feet. This sand, such a nuisance to Grigson on his afternoon spree, gives the sea an extraordinary lustre. The water in the shallow lagoon between St Mary's, Tresco, St Martin's and Bryher can be black and turbid, or sharp as chipped flint, but when the sun comes out it transforms into semi-set transparent green jelly slopping on a dish of white glass. Over the shallowest flats in Tresco channel it turns an unbelievable violet.

Dredge up a wet handful of Scilly sand and it runs through your fingers glittering like half-dissolved sugar. This astonishing stuff leads visitors to compare Scilly to the Bahamas or other idealised, exotic places and its qualities have been prized and cursed in equal measure. Scilly sand is white, palest palomino when wet, and full of sparkling mica. It varies in grade from island to island. St Agnes, the western-most inhabited place, has gritty sand that scrunches underfoot like coarse glass. Tresco's long flanks are heaped with fine powder. Bryher sand has the texture of table salt and St Martin's beaches are packed clean and hard. On St Mary's, sand from Porth Mellon was once considered so superior that parcels of it were sent as presents to the mainland to blot the wet ink on letters and documents. The sand's notable "plenty and brightness" also made it especially suitable for brass polishing and scattering on the floors of houses in Hugh Town.

This beneficent material could turn malignant. In stormy years the fine grains from Porth Cressa Beach are flung against the windows of houses on the Bar with such insistence that they can blast the glass opaque in a single night. In 1822 in *A View of the Present State of the Scilly Islands* the Rev. George Woodley complained: "The fine sand with

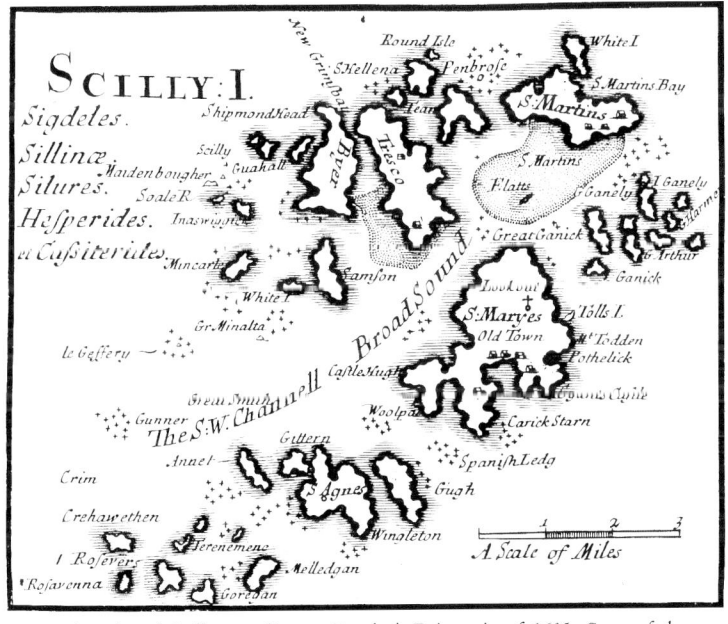

The Isles of Scilly according to Camden's Britannia *of 1695. Some of the spellings have changed, but the proportion and general shape of the islands remains remarkably correct.*

which the roads and many other parts of the Islands are covered is particularly inconvenient. In the summer it dazzles the eyes by its incessant glare: In winter, being furiously blown in the face by the boisterous gales prevalent in that season it stings like a nettle. This sand, and the spray of the sea, are also very injurious to woollen cloaths, especially black, and to hats, which they render brown and rusty in a short time. Most of the Gentlemen on the Islands, therefore, wear straw hats or cloth caps in summer."

For centuries the underwater sands around Scilly provided one of the best tests of position for shipping. Samples hauled up from the sea-bed could be checked and if the captain read the description on his chart properly he could track his position precisely. A navigation chart of 1794 describes the sand north of the great split rock of Men-a-vaur as "like ground wheat with queen shells", while the dangerous area off the western isles has "fine sand pricked with black". There may be more than mica and rock in these samples. Four hundred years ago a ship carrying beads from Venice ran aground on St Agnes and the cargo was washed overboard. The local museum on Scilly displays some of the tiny beads which have since been recovered. Centuries in the sea has roughened and dulled the glass until it is indistinguishable from sand grains.

Out beyond the sheltered inlets of the lagoon a neat line divides the waves into mid-sea blue and shore-sea green; green over the shallow sand, and then deepening into the blue Atlantic. This line is the five-fathom watermark which marks the limit for large shipping on all nautical charts. "Full fathom five" means more than "very deep", it is a border where the land finally cedes its territory to the sea.

St Agnes is adrift outside the five-fathom line and the boat journey from St Mary's to Agnes across the deep is always choppy. Once, coming back at night under a full moon, the boatman stopped mid-way and let the waves rock the launch like a swing-boat so that the moonlight spilt from side-to-side. All around was the dark, strong heave of the sea.

*Blind worker on his donkey on St Mary's.
Picture by John Gibson c1875.*

Archipelagos of self

Seaweed, donkeys and shipwrecks were the features of old Scilly. The famous winter-blooming flowers came later, a Victorian innovation to improve the islanders' miserably low incomes. The donkeys had a hard time of it carrying kelp, stones, salvage and waterlogged dead men. Ore weed was burnt in pits on the beaches and sold for glassmaking. During the kelping season in June and July the islands were shrouded in a pall of mephitic smoke.

It's a great pity there are no donkeys left on Scilly. Their furry robustness and intransigent characters suited the islands. Their tough, boxy hooves coped well with the rocky foreshore, stony lanes and rough headlands stubbed with pine roots. There is a photo of an elderly blind man astride his faithful donkey on St Mary's in 1875. The man's cap is bonneted low over his eyes so that you can't see much of his face. He is smoking a pipe and his feet hang level with the donkey's knees. He gives an impression of absolute immobility. The beast's visage is more expressive. It stands with its forelegs together as neatly as a girl in party shoes and seems to smile.

Tractors have replaced the hard-working Scillonian donkeys, but there are a few riding horses on the islands. These animals are luxuries and they are tended, inevitably, by girls. Why do English girls on small islands want to ride horses? There is almost nowhere that cannot be reached on foot in half an hour. Why do they ride when there is none of

*Scillonian fisherman David Thompson with a catch of dogfish.
Picture by Frank Gibson, early 1970s.*

the social paraphernalia, no competitions, no rosettes to be won, no painted enclosures of showjumps, no-one to pull faces or say "well done," "buck up," or "you're not doing that properly"?

In some shoreline field a girl sets up obstacles on the impossibly springy turf and leaps her horse over them. Higher and higher she raises the rickety pole and there is no one to tell her that she is jumping her own height, that what she is doing is amazing, is dangerous. She would not believe them. There is only the sweating horse, her excited heartbeat and the seawind pouring over them. She is able to be exceptional because she doesn't believe she is.

Islanders set their own standards. They decide what they want to do and then they do it, or not, in their own way, unfettered by notions from other places of how it should be done. What matters is the specific practice here and now on this island where there is no larger comparison with other places. The generality of things is reduced to the specific. On an island with one tree that tree has to stand for the idea of all trees. On this island there are these people, only these, they are the world. How simple, how true, thinks the kimberlin. As long as nobody comes along and says "we do that differently" then the islander has a choice between doing things the way they have always been done, or striking out in a new direction. There is no confusing plurality, things are one thing or the other; follow the custom, or make up your own rules. Mainlanders find this choice romantic because they think it offers the chance to be authentic and creative.

If we weren't deafened and squashed by the external pressures of the modern world we too could be extraordinary, or so we think. And islands seem to offer that space to hear and to expand, to be ourselves. How marvellous to be ourselves!

Gathering kelp for fertiliser on St Martin's in the 1890s. The man was known as Old Ford, the donkey's name is not recorded.
Picture by Alexander Gibson.

Look, stranger

Find a quiet beach on Scilly – perhaps the western side of St Agnes looking out over Smith Sound to the Hell-weathers – and sit down. On a calm day when the waves are not throwing spume the silence is large enough for you to hear the blood moving in your veins. This the sound behind your heartbeat, the same hushed murmur you hear when you put a shell to your ear. Nowadays it's tremendously rare to hear all your internal processes; the click of an eyeball, the squeak of muscle fibres, the whoosh of lungs expanding and contracting, the distant roar of your own private tidal surges. The experience is imprisoning, you realise how body-bound you are, how stranded in the particular, on this island now, in this body. You have reached the island at last and its insularity is fierce, inescapable and ultimately intolerable.

Isolation

From prehistory Scilly has been a place of imprisonment, exile and death.

In Roman times the Emperor Maximus banished heretics and prophets to the isles in order that their prophesises and agitations would be dispersed harmlessly by the Atlantic wind. In the medieval era Scilly was home to reclusive monks and ascetic hermits. Throughout the turbulent seventeenth century Star Castle on St Mary's confined a series of political prisoners too contentious to be held in the Tower of London. Robert Adam's sturdy fortress was intended primarily as a defence against the Spanish, but after it was completed in 1594 its merit as a prison was too convenient to be resisted. Religious fanatics and seditionists were kept there, their activities safely neutralised by thirty miles of rough sea. Among them were Dr John Bastwick, author of inflammatory pamphlets against Archbishop Laud, plus the Duke of Hamilton and seven popish priests. Charles II fled there in 1646 and spent six weeks contemplating his lost kingdom before making a dash for a more civilised exile in France.

There are degrees of exile within Scilly. Before the Victorian lighthouse was built on Bishop Rock legend says that thieves were take out there and left with nothing but a pitcher of water. Escape was impossible. The flat top of the rock was barely above the waves which in stormy weather would sweep over and wash away the felons' bones. An unnamed mother and daughter were said to be the last souls to be marooned there. Bishop Rock was island all right.

For thousands of years Scilly was regarded as a place of death. Most of the high ground is pimpled with ancient graves. Archaeologists argue that these tombs are too numerous and too sophisticated to contain the bones of the indigenous islanders who dwelt modestly in primitive stone villages. Scilly is generally thought to have been some kind of holy burial ground for mainland chiefs and princes. Experts point to the old belief that the spirits of the dead cannot cross water. Yet there must be more to it than that. If quarantine behind water was the only consideration then why not use more easily accessible islets near the coast? Why risk a voyage which all the commentators agree was perilous, a voyage which might take twenty-four or forty-eight hours in an open boat with a dangerous landing when one got there? How many of those death ships failed to navigate the reefs and rocks of Scilly and delivered their cargo to the sea?

The difficulty of the journey was the whole point. We know, we think, that Bronze Age cultures were obsessed with rites of passage. They designed mazes and graves with winding entrances, paths for the soul's flight. Scilly has its own mysterious maze of debatable antiquity which pins down the westernmost edge of St Agnes like a round Celtic brooch on the flying, ragged fringe of the British Isles. New Agers choose to believe the Troy Town Maze is more than three thousand years old, rationalists argue it was built by an

The Troy Town maze on St Agnes in its original state, early 1880s. Nowadays the stones appear more banked because visitors walking the pattern have worn grooves in the turf. Islanders were furious when the maze was re-laid a few years ago without their consent. Fowles and Grigson compare the monument to ancient Viking mazes on Scandinavian islands. These faraway mazes carry similar names, such as Trojeburg or Trojin. In English, "troy town" is an old dialect phrase for a confusing mess or a maze. Grigson thinks the Agnes maze has Viking origins and points out that it is on an island whose name is likely to be a derivation from the Old Norse; Hagni's nes, or Hagni's island. The saint came later.
Picture by John Gibson.

eighteenth century lighthouse keeper. Which makes the truer myth?

Cod-pagans and assorted crystal fondlers would say that the maze is old because older is better, modern life is rubbish and the ancients were wiser than we. They would point to the Mediterranean origins of mazes as evidence of their links with mythological symbolism. Idealists believe that all mazes chart a wonderful journey towards spiritual enlightenment and can take us to a happy place. They don't tend to consider the equally likely proposition some mazes, perhaps this maze, in this position, on this island might lead to a negative end, a dead full stop.

Postulants who enter the Troy Town Maze run the risk of being imprisoned at the centre, marooned in an ankle-twisting tangle of stones and isolated from the free-striding turf of Castella Down.

Now let's fantasise about the unknown lighthouse keeper. Gazing out over the treacherous Western Rocks he watches the shipping come and go, always leaving him in the same place. Bored with his inaction he constructs a maze, an image of the circling waters around Scilly and the narrow, weaving paths boats must follow if they are to land safely. At the centre is his panoptic tower where he may sometimes light a guiding fire, depending that is on whether he can be bothered and whether, as often happened, the islanders have stolen the coals to warm their own scanty hearths. He is an unreliable saviour and he is fed up with doing nothing and going nowhere. The maze is a map of his prison.

Perhaps both stories are true. Perhaps the maze is older than the pedants maintain. Perhaps it disintegrated and became overgrown until one day a bored lighthouse keeper mooching about on the clifftop and gazing enviously across the Atlantic stubbed his toe on a stone, looked down and saw a pattern.

I want to get away

Holidaymakers with jobs and kids wash-up on Scilly like smashed boats, their minds, emotions and bodies bobbing in separate pieces. All the wreckage is thrown back together and for the first few days they rush about from island to island on frantic itineraries creating their own cloud of noise and exhaustion. Have they done Piper's Hole yet? Been on the Seabird Special? Caught a boat trip to Bishop Rock Lighthouse? Visited the museum? Played on the right beaches? Admired the Abbey Gardens on Tresco? At least the madness preserves them from mania. As long as they are charging about trailing guide books and cameras there is no danger they will peer into the islands' mirror and see themselves reflected back in individual singularity.

We spend our lives in distraction and dream that if we could only free ourselves we would discover the wonderful possibilities of our inner islands. Modern people believe that to be "in touch with yourself" is, ergo, to be "in harmony". Creativity is constructive, it helps you "come to terms" with problems and makes you happy. Writing, painting and solitude are said to fire one's energies. The freed self plunges into a kind of mythic whirlpool bath which soothes aching emotions. You emerge radiant, better, truer, realer and more alive. Give yourself a holiday, slip into something more comfortable, go on, treat yourself, it's your lifestyle.

Spiritual day-trippers taking a quick excursion to the soul's islands often find that they are as ill-prepared as the tourists who arrive on the *Scillonian III* wearing high heels and thin cardies. What if we want, like, need to be distracted and fragmented rather than condensed, what if solitude and creativity are not therapeutic but damaging and disturbing?

Old Church St Marys

REACHING THE ISLANDS

Waiting to go to islands,
Thinking of mountains,
Listening to birdsong.

The sea lacks birdsong
On the way to islands
And the waves will be mountains.

Only the mainland has mountains
But Scilly has birdsong,
The pleasure of islands.

And here in the islands
I can see distant mountains
And am showered with birdsong.

Being an islander

If you live on an island everyone will know if you fail. You have set a course and if your ship goes down there will be no disguise and nowhere else to turn. There will be no pretending. Success on an island has the quality of heroism on the mainland.

"I've never asked for a job, never been in want of work and never got the sack," says Matt Lethbridge, Scilly's greatest living islander. Three generations of Lethbridges were coxswains on the Scilly Lifeboat over a period spanning seventy-one years. "Young" Matt Lethbridge, whose father was also called Matt, was on the Lifeboat crew for nearly forty years and coxswain for nearly thirty of them. He was also a fisherman and in winter-time he worked with carpenters, blacksmiths, painter and decorators and in a butcher's shop. During the Second World War he joined the RAF Air-Sea Rescue Service. He was lent to the Navy and stationed on Scilly, but he fell out with an officer and got posted overseas to Africa. The Navy was extremely annoyed. He was so reliable and so "on top of his job" that a squabble ensued between the Navy and the RAF as to who should have the use of him. In 1982 Eammon Andrews devoted an episode of *This Is Your Life* to Lethbridge, who was tricked into going to London and ambushed at the Boat Show. "Not my thing, all those fancy yachts," he complained. The programme makers reunited him with many of the people he had saved from the sea, including some

Matt Lethbridge. He stopped smoking soon after Frank Gibson took this picture in the late 1960s.

survivors of the helicopter accident which happened off St Mary's the year before.

Now in his seventies, Lethbridge can look back on a life of unusual independence and diversity. A studio portrait taken by Frank Gibson shows a handsome man confident enough to turn his dashing pose into a small joke. He looks the kind of person who could teach himself to do anything.

"We got on with the job – you were free from apprenticeships, you learned from others and did what you wanted," he says with the self-reliance characteristic of islanders.

Lethbridge's great-great-grandfather was one of the stonemasons who came out to Scilly to build the Bishop Rock Lighthouse. He married a Samson girl and settled on St Mary's. His bride, Ann Woodcock, belonged to one of Scilly's oldest families and through her his descendants could claim kinship to all the chief Scillonian clans. These connections reach further than might be expected. When Matt Lethbridge's father fell in love with the daughter of a St Ives fisherman, the prospective father-in-law made enquiries among his own relatives on Scilly and found that the two families were already related through the Nances of Tean and St Martin's.

In 1985 Lethbridge retired from the Lifeboat. In his spare time he turned his hand to a new skill – painting. He had always been interested in drawing and occasionally produced sketches of boats. Now he sat down to teach himself how to paint. Using oils or acrylic and working on board he depicted the shipping history of Scilly from the tall ships like the *Cutty Sark* in which his grandfather once sailed, to the Lifeboats and fishing trawlers which plied the local waters. All his pictures show boats at sea, mostly in rough conditions. Great care and skill is expended in painting the waves to capture their glassy, heaving strength. The skies are atmospheric, sometimes grey and lowering. Frequently he includes a backdrop of island shoreline. The ships are taken from photographs or illustrations but the passing landscape he paints accurately entirely from memory. The boats and the shoreline are foils for the main subject, the sea. No figures or animals appear, these are elemental studies of the manmade versus the natural without the intervention of personality. "I try to paint the wind and the sea," he explains.

Lethbridge's pictures are proper pieces of art, attempts to depict an experience rather than prettify a scene. The boats carve into the sea, cutting through waves which threaten to overwhelm them. Free from the deliberate charm and naivete of amateur painting these works present a stern view of the sea and its perils. By eschewing weedy watercolour pastels and sunlit scenes Lethbridge has chosen to paint Scilly as the islanders know it. Strong colours, dark greens and blues, convey tangled emotions. Many of these pictures show ships in distress or the Lifeboat on the way to a particularly difficult rescue.

Lethbridge's artistic talent is a Scillonian trait that runs through certain island families. His Uncle George, who was drowned out fishing a few months after returning safe from the First World War, was similarly gifted. Beyond this there is a more distant connection to the Lethbridge family, which suggests that Scillonian blood has already produced a great artist.

Sketch of a fishing trawler by Matt Lethbridge. Named after Lethbridge's daughters, this is his ideal imagined boat. His real boat was smaller, not so well-equipped and called the JML *after his Uncle Jim and his father, Matt.*

In view of Scilly

In the Tate Gallery in St Ives hang the paintings of a second fisherman who took up art late in life. Alfred Wallis' primitive works of ships and houses are not to every taste. They defy the conventions of perspective and use a sombre, limited palette. In Wallis' day local people in St Ives regarded them as the pathetic daublings of a senile old man. Only a clique of London writers and artists who visited St Ives in the 1920s and 30s recognised their true value. Ben Nicholson and his circle bought paintings for a few pence — one cannot help feeling that they diddled the old man. Wallis died in the workhouse in 1942 maddened by poverty and age (he was eighty-seven and penniless). He has since been recognised as one of the twentieth century's most important British painters.

Wallis' parents were married in Penzance but he was born in Devonport in Plymouth in 1855. When Wallis was ten his mother, Jane, died and the family returned to West Cornwall. Wallis always said that his mother came from Scilly and until recently books about Wallis took him at his word. Then it was discovered that parish records show Jane was baptised in the tiny village of Sennen near Land's End. In his study of Wallis, Matthew Gale, a curator at the Tate, ignores the Scillonian connection entirely.

Jane was born an Ellis. On Scilly, Ellis is one of the oldest family names, a distinction shared with Woodcock and Nance. While the Woodcocks came from Samson, the

*St Martin's, the most remote and self-sufficient of the islands.
Picture by Frank Gibson, 1960s.*

name Ellis is associated with St Martin's which has long been regarded as the most insular and independent of the Off Islands. No-one knows why St Martin's men were nicknamed "Ginnicks", but they were said to be tall and thin with red or sandy hair. Moreover, they were believed to have come originally from Sennen. During the eighteenth and nineteenth centuries there was constant to-ing and fro-ing between Sennen and St Martin's. People from Sennen went out to Scilly to fish and many had relatives on St Martin's.

It is inconceivable that Jane was not related to the Ellises of St Martin's. Sennen was too small and the links were too close. But why did Wallis insist she was a Scillonian? She might have spent her childhood with relatives on St Martin's. It could even be possible that her mother was in fact a St Martin's woman who chose to return to her family in Sennen for her lying-in. Childbirth was fraught with hazards and women often returned to their families if they could expect more care and attention than they would get at home. It's not unreasonable to think that Jane's mother opted to give birth on the mainland where help was more readily available than on a notoriously remote island. If she did, it would explain why Wallis always insisted that Jane was a Scillonian although she was baptised in Cornwall.

Go back far enough on Scilly and the connections between families become as intricate and baffling as the ways through the reefs and shoals around the islands. Everyone is related to everyone else and the net of relationships stretches over the sea to Cornwall and back again – as Matt

Lethbridge's maternal grandfather discovered when he inquired into his son-in-law's antecedents. It's likely that the Lethbridges are related to the Ellises through the Woodcocks and that Matt Lethbridge's artistic talents originate in the same gene pool as Alfred Wallis'. When Wallis took up painting "for company" and stuck to his work despite the ridicule of his neighbours, he showed his island legacy for independence and self-reliance.

This same pioneering, artistic spirit animated another St Martin's family and led to the founding of a unique Scillonian archive. John Gibson was a third seaman who developed an interest in art. In Gibson's case the urge to create found expression in photography, the newest art form of his day.

So that strangers might see

Five generations of the Gibson family have photographed Scilly during the last one hundred and thirty-five years. The remarkable archive they created covers all aspects of island life; shipwrecks, industry, flower farming, public events and private domestic scenes. Alexander Gibson, son of John, explained that the purpose of the collection was to produce "a series of photographs so that strangers might see what wonders this little known archipelago held".

John Gibson, the first photographer, was born in 1827 on the Aran Islands off Ireland where his father was a coastguard. His parents came from St Martin's and after his father died in 1840 the family moved back to Scilly and opened a shop on St Mary's. John was a sailor and there is a story that he bought a camera on a voyage to China. He tinkered around with cameras in the late 1850s, learning more about his craft from two photographers in Penzance. In about 1860 he moved to Penzance with his wife Sarah and they opened a studio on the Promenade specialising in portraits, local scenes and events. The family kept on the shop in St Mary's and by the turn of the century they had a studio in Church Street.

John was in his mid-thirties when he set up the Penzance studio – already middle-aged in nineteenth century terms. By then he had two sons, Alexander and Herbert, and in 1868 Alexander joined the business at the age of twelve. Herbert was soon to follow and for more than thirty years the

Alexander Gibson.

trio worked together producing some of the finest images of the period and becoming renowned for their shipwrecks. At first they used the wet plate Collodion process which required cumbersome developing equipment to be carted around with the photographer. Plates had to be wrapped in a cloth and developed within the hour. It must have been a relief when the slightly more portable dry plate method came in after 1871.

Both processes allowed smudges and blurs to be painted out afterwards and Alexander in particular exploited the

Frank Gibson.

opportunity to touch-up his prints in order to make a better picture. His aim was artistic rather than corrective and he went to a lot of trouble to enhance atmospheric effects. His artistic sympathies were united with strong journalistic instincts and if he missed a particular shipwreck he wasn't above painting it and selling his photograph of the painting as the real thing. According to Rex Cowan's fascinating history of the Gibsons in *A Century of Images*, Alexander actually trained as a painter at the Penzance School of Art. Genuine talent and the islander's impulse to fix things sometimes made him over-eager to improve his pictures, but his best compositions are remarkable.

Alexander was an eccentric. He wore a velvet cap, nurtured an obsessive interest in Druidism and grew a beard which straggled over his breast like white sea foam. He was tall with long hair and a voice that clacked like a gull. He quarrelled furiously with his son James, who entered the business in 1916. At the end of the 1930s an enraged Alexander smashed hundreds of the original photographic plates, destroying years of work, and then retired to Shropshire. Shortly before he died in 1944 he was elected a Bard of Cornwall and given the name "Tas en Enesow" or Father of the Islands.

James' son Frank was born in 1929 and took over the firm in 1958. He has kept Scilly in the news and built up the trade by introducing modern methods and colour process. Now semi-retired, the studio is in the hands of his youngest daughter Sandra Gibson, the fifth generation of Gibson and the first woman to run the business.

The leaping light

You see things in a different light on Scilly. The light burns you, literally. It eats through the strongest sunblock to redden exposed skin and brings the fair out in dog-blotch freckles. Trippers on the *Scillonian* are warned about the strength of Scilly sunlight and advised to cover up. This clean, scouring brightness belongs to the sea; it's an offshore quantity, not like the easy, appealing light of West Cornwall. It entranced the painter Patrick Heron who called it "the whitest light in Europe". When you first arrive on Scilly your eyes are still used to the thick mainland air and everything appears bleached and faded like the pattern on a bathing towel that has been exposed to too many summers' salt and sun. After a few hours your vision adjusts and the colours leap back at you clean and bright as if the grime has been filtered out of them, swept away by the pollution-free Atlantic wind.

"One has only to spend a few hours on the Isles of Scilly to be virtually brain-washed by a very special kind of physical space and remoteness," said Heron.

Economies of scale

Everyone who goes to Scilly is struck by its dizzying lack of scale. The islands are toy humps in the ocean and the main island, St Mary's, is only two and a half miles long. Yet if you walk around any of the isles it is impossible to judge whether landmarks are a few yards or a few miles away and whether the cliffs tower thousands of feet or drop a few hundred inches. Clambering about on rocks on the beach you lose your sense of perspective and vertigo may set in leaving you clinging to a ledge as if there were a chasm below instead of a short step that a terrier dog could bound down in safety. Sometimes your view oscillates so fast between the vast and the tiny that you become paralysed. Time bends and nearby places take longer to reach than you think, while those apparently far away are brought close in seconds.

Paying intimate attention to the size of things makes you realise how small and precious it all is. Every footfall on headland turf crushes an individual plant. Stick to the narrow, worn paths and you feel enraged that other careless walkers have squished the hairy brown caterpillars and iridescent beetles beneath their clodhopping boots. There is a strong feeling of irreplaceable things being used up and destroyed forever.

Previous generations were free from this ecological neurasthenia. For the inhabitants the islands were big and robust enough to be whole countries. What else can account for the re-use of place names? You'd think it would be confusing to have two Piper's Holes, and two White Islands within the archipelago. The answer must lie with the inhabitants' extreme insularity. Why should St Martin's people cause confusion when they referred to White Island if they never visited its namesake the other side of Bryher? For people on Tresco, Piper's Hole on St Mary's was mentally as far away as Penzance. It is a sign of how far our horizons have widened that we find this attitude at all remarkable. Until the second half of the twentieth century it was quite normal not to travel beyond the place you were born, look at the Portlanders who never ventured to the next village a mere three miles away.

The Master

In *The Master* by T.H. White, two children are stranded on the tiny island of Rockall at the mercy of a madman who is building a machine to take over the world. As far as the children are concerned, the world has shrunk to Rockall and the sinister Master is already in full control. *The Master* was published in 1957 at the height of the Cold War and its literary legacy lurks unsettlingly round the corners of another tale of an island with a manipulative madman. John Fowles' tricky novel *The Magus* (first published in 1966) features a recluse who plays teasing sexual games with a young initiate. Behind the ruthless, egotistical figures of the Master and the Magus stand a legion of islomanes, real and imagined, from Robinson Crusoe to Prospero. All of them are the inventors and despotic rulers of their own kingdoms.

In 1834 Scilly was leased to Augustus Smith, a landed gentleman from Hertfordshire, whose genteel upbringing might have fitted him for a part in a Jane Austen novel. Like so many gentlemen of his time he really had nothing to do. For a while he busied himself with schemes for educational improvement in the country around Berkhamsted but this wasn't enough and there was a strong feeling that he was a young man in Need of an Occupation.

Scilly was (and still is) the property of the Duchy of Cornwall. In the early nineteenth century it was a desolate place, wild and uncultivated, reliant on mainland charity. The land was too bare and windy for farming and people on

Augustus Smith poses with the St Agnes pilots in 1872 on the island where youths once tried to murder him. By this late stage in his life the "old Caliban" was treated with more deference. The pilots were men who knew the reefs and shoals around Scilly and guided ships around their treacherous shores. Picture by John Gibson.

the Off Islands often faced starvation. Smuggling supplemented by wrecking was said to be the islanders' only source of income. The miseries of the inhabitants were depicted in a plaintive succession of reports, which drew public attention to the islands' poverty. The lease on this pitiful domain had fallen to the Duke of Leeds and in 1831 it expired. Scilly was a masterless ship adrift in the ocean. After some years of wrangling, Mr Smith took over the lease and with it the glorious title of Lord Proprietor of the Isles of Scilly.

The new Lord Proprietor believed in the justness of his vision. His approach to Scilly was deeply feudal and he treated the land and the people as a ruthless private experiment in social improvement. He strove to create a tidy landscape of productive fields and neat homesteads occupied by respectful tenants with orthodox Christian minds. Those who refused to fit in with his policies, who were slovenly, untidy, irreligious or over-fond of drink, were banished to the mainland. In the early years of his rule the number of paupers arriving at Penzance from Scilly cost the authorities considerable embarrassment. In the first twenty years of Smith's rule the overall population of Scilly increased by 141 to 2,606 as a result of a churning process which caused great upheavals among resident families. Stories of how Smith revived the islanders' fortunes and made them stand on their own feet have to be balanced against his attitude to the Scillonians who were living there when he first arrived. The Lord Proprietor was determined that in his kingdom the poor would *not* always be with him.

Before settling on Scilly, Smith investigated several estates in Ireland as possible subjects for his improvement schemes. One suspects that there was never any real question in his mind as to which location he preferred. Shifting the recalcitrant Irish over an arbitrary border from one parish to the next would have been a tougher challenge than bundling destitute islanders on to a ship.

Everyone who visits small islands dreams about being in charge of a private kingdom. Smith lived in an era when imperial fantasies were considered right and proper among men of a certain class. Furthermore, they were expected to turn their daydreams into action. British men strove to spread "civilisation" and progress across the globe. A fine example of this was Captain John Clunies-Ross, who settled on the Cocos (Keeling) Islands in the Indian Ocean a few years before Smith took over the lease on Scilly.

All twenty-seven islands in the Cocos group were uninhabited, but Clunies-Ross wasn't alone. There were his Malay servants, plus the unwelcome presence of Alexander Hare, his former friend and shipping partner. Hare settled on a neighbouring island and claimed it as his own. A long, territorial tussle ensued between the two men. Hare was a notorious figure, more pirate than trader, whose behaviour in the East Indies was an exaggerated, swashbuckling parody of colonialism. He was the first Englishman to be given the title of White Rajah and he was rumoured to keep a harem. His numerous children ran naked on Prison Island until they were old enough to work in the new cocoa plantations.

Clunies-Ross by contrast was an upright Presbyterian whose family could trace its descent back to the Scottish kings. He was proud of his lineage and called his settlement New Selma after the ancient, legendary capital of Scotland. With the aid of convict labour he worked hard to strip the atolls of their natural flora and establish cocoa plantations, extinguishing the indigenous wildlife. The workers supplemented their rations with birds, fish, turtle eggs and coconut crabs. Imported rats and cats ate anything the humans didn't fancy and the southern atoll became a silent, sterile coconut monoculture. In 1836 Clunies-Ross built a splendid residence, entirely constructed from stone shipped out from Scotland, which boasted a ballroom, a billiards room and ten bedrooms.

For seven years Hare and Clunies-Ross lived as rival kings. As Clunies-Ross implemented his industrious schemes, Hare grew ever more eccentric. In 1834, coincidentally the year Augustus Smith came to Scilly, Hare disappeared. Thereafter, Clunies-Ross ruled alone, founding a dynasty that survived until the 1970s when the enormous family fortune was lost in a disastrous shipping venture and the archipelago became Australian territory.

Hare was English, Clunies-Ross Scottish. More significantly, Hare was a mainlander; Clunies-Ross was already an islander by descent. His family came from the bleak environment of the Shetland Isles where his ancestors were infamous for their regimes of extortion and oppression. Lawrence Bruce, who came to Shetland in 1571, was particularly hated and feared. Bruce was closely related to James V of Scotland and may have been the king's half-brother. He was deprived of his office of "underfowde" after the islanders petitioned the Privy Council in protest at the harshness of his reign, but he retained his lands and continued to "live a life of constant disturbance and interference with the rights of others" until his death at the age of nearly eighty.

The impulse to control and improve was part of John Clunies-Ross' heritage. His father ran one of the first schools set up by the Society for the Promotion of Christian Knowledge, a movement originally founded with the intention of Anglicising and civilising the Gaelic highlands. By the turn of the nineteenth century SPCK had widened its remit to other independent, ungodly regions and had organised several missions to Scilly.

Like Hare and Clunies-Ross, Augustus Smith enjoyed the liberty of a private kingdom situated outside the social boundaries of the civilised world. Smith enjoyed the delightful freedom of Scilly where there was no "society". Rumours about his conduct were unlikely to percolate as far as London where he still spent several months a year carrying out his duties – from 1857 to 1865 he was MP for Truro. When on Scilly he chose to live on Tresco so as to have an island to himself (the peasants didn't count). He built a house on the ruins of the abbey with commanding views over the islands. His private rooms had a secret door for admitting lovers and when he died he left legacies to two Tresco women who were said to have borne his children.

When the Lord Proprietor's friends and relations visited there was plenty to entertain them. They went sightseeing and picnicked on Samson, conveniently purged of its inhabitants. They shot wildfowl and the rare birds which flew in exhausted from Africa and Europe. In 1849 Augustus Pechell "was very well amused" to kill a wryneck, a reed warbler and an osprey. (The hundreds of fanatical bird watchers who descend on the isles today would spit feathers.) Smith's fabulous sub-tropical gardens were an amusing novelty, a fantasy landscape which remade the island and replaced indigenous flora with exotic foreign plants.

It's tempting to make Smith out as a monster, but that would be too simple. He provided schools for all Scillonian children and fined parents if their offspring didn't attend; this was years before education became compulsory on the mainland. He believed fervently in the virtue of self-reliance and he was an enthusiastic distributor of Samuel Smiles classic Victorian manual *Self-Help*. Before coming to Scilly, Smith set out what he saw as "the great ends of education" in a pamphlet, *An Apology for Parochial Education on Comprehensive Principles*. The purpose of education was "to teach a man his duty towards God and his duty towards man: to teach a man so to live in his world as many render him the fitter for that life to come, and in the meantime to conduct himself in this life as may render him a useful member of that Society amongst whom his lot is cast."

A stroll in Tresco gardens was one of the genteel delights available to guests and visitors in 1883.

It's ironic that by teaching the young Scillonians to read Smith inspired them to nickname him "that old Caliban". Many islanders hated him. They broke his windows, cast spells on him and a gang of youths on St Agnes once rolled him up in a sail and left him on a beach at the mercy of the incoming tide. His reign was not universally approved outside Scilly either. His friends dubbed him The Emperor, the islanders called him The Governor, and Londoners dismissed him with a sneering pun as Scilly Smith. The philosopher John Stuart Mill was scathing about Smith's achievements: "There is prosperity of a kind, undoubtedly," he admitted, "but it is paternal government. I detest paternal government."

Home rule

There may be something about the very nature of island ownership that turns masters into autocrats. The bare bones of despotism wear through the ragged trousers of philanthropy soon enough.

If you were to sail away from Scilly in an insanely straight line across the Atlantic and the wide Sargasso Sea, the first land you would encounter on the other side would be a small cay in the Bahamas. Sixty years ago Whale Cay (pronounced "key") was the home of an eccentrically mannish heiress who built a big house, "reformed" the natives and hacked herself a kingdom out of the scrub. Put Marion Carstairs alongside Augustus Smith and you have a perfect anatomy of islomania.

The story of "Joe" Carstairs' extraordinary life is told by Kate Summerscale in *The Queen of Whale Cay*. Carstairs was a speedboat-racing lesbian who dressed like a man, had numerous lovers and seduced Marlene Dietrich. In 1934 she bought Whale Cay and set about making its 500 islanders properly self-sufficient. Carstairs emphasised the island nature of her kingdom and strove to make it appear more remote by telling everyone it was at the edge of the Atlantic.

In the process of refashioning the island Carstairs remade herself: "The persona she constructed drew on old-fashioned models of manhood: she stood for Empire, Britishness, cleanliness, hard work, physical bravery, moral fibre," says

Summerscale. Culturally if not sexually, Carstairs and Smith were cut from the same cloth.

Like the Off Islands of Scilly, the Out Islands of the Bahamas were resistant to cultivation. The inhabitants followed the usual old-fashioned island lifestyle; living off wrecking and piracy, boozing when they could, starving occasionally, practising magic and fornicating when they felt like it. Like Smith before her, Carstairs set out to tidy things up, to teach her subjects "to live in cleanliness and order so that they and the generations of the future, shall be fitted to make a decent livelihood, and to be an asset rather than a liability to the community." Roads, religion, education and the means to earn an honest living were the way forward. The unreformed were horsewhipped or banished.

Carstairs enjoyed the freedom of behaving as she pleased. She saw nothing odd in pursuing exotic liaisons away from the wagging tongues of society while at the same time enforcing restrictive morals on her islanders. She commanded her subjects to live chastely, preferably segregated by sex, while she entertained a string of girlfriends behind the private walls of her luxury mansion.

Both Carstairs and Smith used the influence of religion to break the islanders of their old habits. They faced similar problems in dealing with independent people whose Christian beliefs were erratic. They realised that orthodoxy would supply clear social rules and they built new churches and imported Anglican curates to preach in them.

When it came to religious matters Carstairs was undoubtedly the more straightforwardly cynical. She insisted her subjects attend church, but blithely employed a decadent priest who made sexual advances to choirboys. The righteous Smith was genuinely concerned by the religious confusion he detected among the Scillonians.

Scilly was fissured along class lines into Anglicans, Wesleyan Methodists and Bible Christians. Most of the poorest Scillonians were Bible Christians or Bryanites, a breakaway Methodist sect which converted many of the working class in Cornwall and Devon. They built meeting houses on St Mary's, St Martin's and St Agnes and had a woman minister who conducted services in the islanders' homes. Women were admitted to the ministry equally with men and members of the congregation sometimes performed the last rites. Burials and christenings were seldom registered correctly. All this was very worrying for a man of sincere belief who was obliged by the terms of his lease to strengthen the work of the established church on Scilly. Lack of accurate registration of births and deaths also happened to be extremely inconvenient for anybody who wanted to keep a close eye on the population's comings and goings.

There was another, evasive element to the islanders' beliefs. Witchcraft and magic persisted on Scilly. In the Bahamas there was obeah, a variant of voodoo. Belief in magic need not be anything to do with religion. Its secret practices tend to flourish among people who lack economic power as a way to gain advantage over social superiors. In

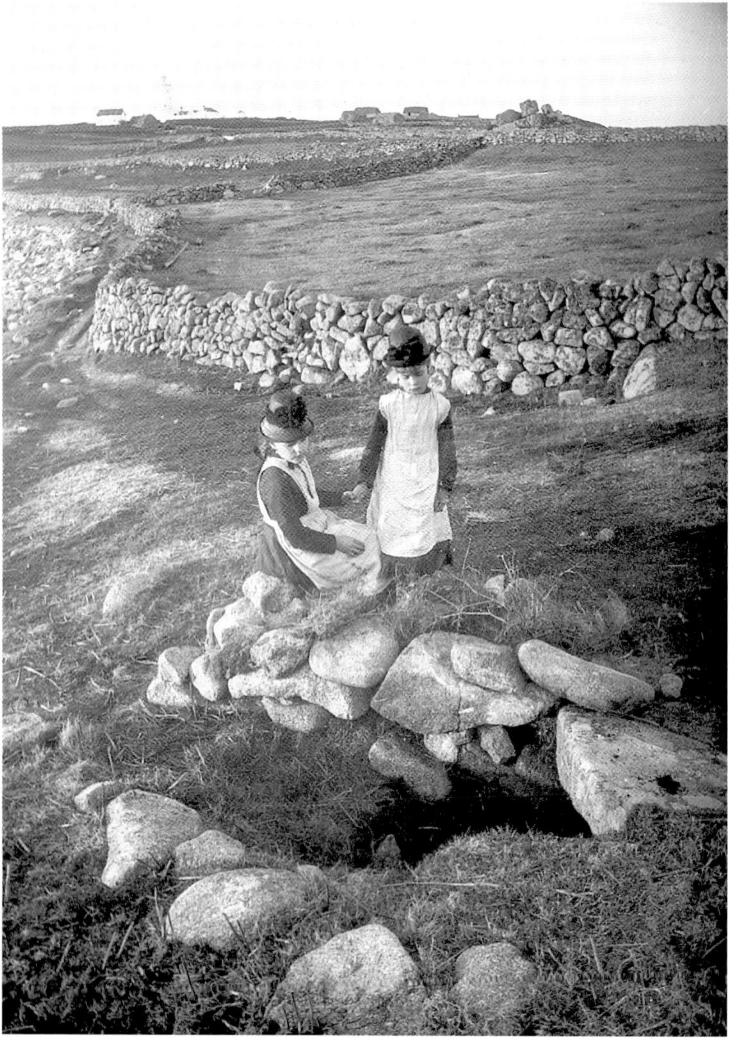

patriarchal, Christianised societies which disapprove of magic, most of its practitioners are female. Carstairs and Smith both had run-ins with native methods of getting even.

Tresco was said to swarm with witches and Samson was a stronghold of witchcraft until Smith evicted all the inhabitants in 1855. (Matt Lethbridge's great-great-grandmother was twelve at the time). It was on Samson, according to his biographer Elisabeth Inglis-Jones, that Smith was spellbound by an old crone. "As he was about to get into the boat he suddenly found himself unable to stir. His legs, heavy as lead, rooted him to the shore. Not one inch could he budge whatever his men did to help him, pull and push as they might." He was released only when the boatmen persuaded the woman to lift the curse.

Being less conventional than Smith and, crucially, a woman, Carstairs was able to turn her islanders' belief to her own advantage. She had a man doll she carried around with her as if it was a fetish and most importantly she claimed the ability to heal. Her subjects were captivated.

Opposite: These little moppets gazing into St Warna's Well on St Agnes were the subject of a popular Victorian postcard. The spooky atmosphere is more than coincidental — this ancient holy well was dedicated to a Celtic saint who was believed to be instrumental in providing wrecks. When the islanders wanted a wreck it was said they would go to the well, mutter a prayer and drop pins into the water. Scores of ships have struck the Western Rocks since St Warna arrived in her coracle from Ireland.
Picture by Alexander Gibson, mid-1890s

If Smith had been able to annex similar powers he might have met with less suspicion and resistance from the islanders. Female healers had a long and distinguished tradition on Scilly. Writing in 1750, Robert Heath described how sick islanders were treated by a select group of knowledgeable women. He called this unnamed body "a Society of skilful Aunts". (Heath says that all Scillonian adults of a certain status were accorded the title of Aunt or Uncle as a mark of respect.) Members were especially skilled at midwifery and used a number of secret remedies whose recipes had been passed down through the society for hundreds of years. Their venerable leader or president was one Aunt Sarah who was notable for her long beard. Stroking Aunt Sarah's beard was believed to induce miraculous cures.

Heath regarded the women with a mixture of admiration and condescension characteristic of his time. He had difficulty describing their medical vocation and resorted to approximations and diminutives; "a Sort of College of *physicians*... the attending *Doctress*". He made clear that the Society came into being "for Want of *Male* practitioners in Physic". In praising the Society he dwelt less on its knowledge than its good sense for not interfering as Nature worked its cure. He used the women as an example to make satirical comments at the expense of the established (male) medical profession. The overall picture is of a quaint and slightly comical group of old biddies doing their best and with God's help achieving more than might be expected. Nevertheless, Heath's jocular style can't quite suppress the possibility that Sarah Jenkins and her followers could have possessed an unusual level of genuine skill.

Of course one has to be careful about imposing modern notions on people of the past, but there is enough evidence to speculate that Smith antagonised the islanders because he cut across a long-established network of social identity. Carstairs by contrast did many of the same things but achieved popular support by tapping in to her subjects' beliefs.

One of Scilly's problems was the lack of land suitable for cultivation. Families were bunched on plots far too small and unproductive to feed them. The situation was made worse by the island tradition of dividing farms so that all children, including daughters, inherited a piece of land. The farms shrank and families starved. Smith had the common sense to insist that only one child – assumed without debate to be the eldest son – could inherit. Younger children and girls would have to find work elsewhere, even move to the mainland if need be. It probably never occurred to Smith that he was disinheriting half the population; after all females on the mainland didn't inherit of their own right. His ruling caused great aggravation and ill feeling. These were islands with a proud dissenting spirit that did things their own way. Here, women could hold land and some of them were valued for their esoteric knowledge. Smith's reforms involved more than land; he re-designed the way the islanders thought about themselves.

It would be a crashing anachronism to suggest that the islanders consciously framed their objections to Smith as protests against patriarchal oppression. The Scillonians were more concerned about the immediate survival of their families and whether Smith was going to build a new harbour pier in St Mary's Pool (he did and it helped the islands create prosperity through shipbuilding). But the ways that resentment leaked out show the sources of the hurt. No wonder the feeble old crone of Samson was able to bind the mighty Lord Proprietor – he knew in his subconscious that he was forever in debt and could not be released without her pardon.

Little Sarah Pender, who was famous for her small stature and extra fingers and toes — she had five fingers and a thumb on each hand and six toes on each foot. She lived with her mother Aunt Sarah on Bryher. Rex Cowan notes that the name Aunt Sarah had mystic significance on Scilly. Both women died on the same day in 1944 — Little Sarah was seventy-three and Aunt Sarah ninety-eight.
Picture by Alexander Gibson, early 1900s.

The Islands compared to a Feast

Holiday isles appeal to the sensual. They delight the eye, they free the mind and rest the body. They should supply exotic tastes; the bounteous, sweet tang of coconut and the swooning burn of rum.

Dried, salted ling, hard and parched as old, cracked sailcloth, and known as "tow rag" was Scilly's greatest delicacy. Scillonian dried ling was famous – there is a theory that Scilly takes it name from "zilli", the Cornish name for the fish. These thin, spare members of the cod family were abundant in western waters around the British Isles. Their long, slender bodies of mottled bronze grew to a maximum length of two metres and on their chins they carried a barbel, a soft, trailing protrusion like a sodden hangnail.

Ling were laid out on the islands' lacy stone walls or "hedges" to dry in the desiccating light. Then their tough, chewy bodies were salted down and sold to sailors. Admiral Lord Nelson "much enjoyed" a parcel of dried ling sent to him in Toulon in 1803.

Ling was the best native food Scilly had to offer. A commoner, everyday dish on islanders' tables was scad or horse mackerel which was caught in large quantities off St Agnes. Scad have big heads, enormous eyes, and a row of bony plates called scutes protecting their sensitive lateral line. Fishermen in Norfolk in the seventeenth century called them simply "horse", and the meaning probably has the old sense of coarse or large. Their bony, woolly, bland flesh must have made an uninspiring dish eaten every day with hard, sandy Scillonian potatoes.

There was worse, however. Archaeological evidence shows that for thousands of years limpets were the base-line subsistence food of Scilly. The Cornish name for limpet is crogan, a word which may be related to the Old English *cróh* meaning small vessel or pitcher of water. Today, when fashionable seafood chefs revive provincial recipes for rubbery, sewagey razor shells and the snot trombones of whelks, nobody eats limpets. There is a good reason; limpets make the vilest, meanest of foods.

Every childhood seaside holiday has its casual cruelties. Sand stuffed in the mouth or down the pants, bored adults who break their promises, the fear of being cut off by the tide with evening coming on; and kicking limpets off slippery rocks. Poor limpets, skittering and bouncing away from sandalled heels, some to be smashed with a beach pebble, and then to dip your finger in the sticky, oozy salt, a disgusting taste.

Limpets are easy to harvest and can be eaten raw, though they are slightly less uneatable when cooked. Matt Lethbridge remembers boiling them up in an old biscuit tin of seawater over a driftwood fire on the beach. At Herm in the Channel Islands they were covered with heaps of lighted straw and immolated in their shells. In Normandy they were placed upside down on a hot gridiron until their water evaporated. Scottish islanders used to mix limpet juice with oatmeal canny people don't eat the actual mollusc. In his seminal work *North Atlantic Seafood*, Alan Davidson recommends boiling limpets for about half an hour so that they yield a rich, brown liquor. He advises drinking the liquor and discarding the limpets' shrivelled carcasses. At times of famine in the early nineteenth century whole families survived on nothing but limpets bashed off the rocks. The 1818 *Report detailing the Extreme Miseries of the Off-Islands of Scilly* painted a gruesome picture of a people reduced below prehistoric competence. In 1822 the Rev. George Woodley reported that limpets were often fed to pigs "which gives the flesh a disagreeable redness, and a very unpalatable taste."

As times improved limpet eating retreated into tradition. According to Jessie Mothersole, it became a Scillonian custom to gather limpets on Good Friday, traditionally a fast day. In her book *The Isles of Scilly*, published in 1910, Mothersole recounts the words of a limpet-eater who seems to have consumed the mollusc raw and alive. He said it was "tough and elastic, like a piece of india rubber, and if you don't mind, when you're trying to bite it, it will fly back in your face and give you a black eye."

Modern visitors to Scilly may be thankful that their worst culinary trial is enduring a week of eating frozen, processed chips with every meal. Provisioning the islands has always been a problem. When G.H. Lewes and George Eliot spent seven weeks on St Mary's in the spring of 1857 Lewes was highly disgruntled by the limited menu.

"Beef *is* obtainable – by forethought and stratagem; but mutton is a myth... Poultry, too, may be had – at Penzance;

George Symons guts a pollock. Apart from Matt Lethbridge, George was the only other member of the Lifeboat crew who was said never to be seasick. Picture by Frank Gibson, 1988.

and fish — when the weather is calm, which it never is at this season."

Lewes was disgusted by the spectacle of the vegetable cart that visited Hugh Town twice a week to hawk its goods around the streets. "If you like to gather round it, as the cats and dogs do round the London cat's-meat-man, you may stock yourself with vegetables."

Lewes could be seen as a forerunner of some of the modern visitors to Scilly; middle-aged in years but already elderly in habit, probably rather red-faced, thinking continually of his belly, unwilling to disrupt his routine and annoyed that the islands didn't come up to scratch. Not all of his scorn is meant to be taken seriously, there is heavy, Victorian humour and cumbersome self-parody in his grouchiness.

"Spiritual-minded persons, indifferent to mutton, may disregard this carnal inconvenience, and take refuge in the more ideal elements of picturesqueness, solitude, and simplicity. I cannot say that the inconvenience weighed heavily in the scale against the charms of Scilly; the more so, as an enlarged experience proved the case not to be *quite* so bad as it seemed at first."

There were other ways of catching fish than by going out in a boat. Rockpool blennies could be enticed out of the kinches between the stone with a piece of limpet on a pin. A mysterious type of shoreline fish could be seduced into taking the bait if the angler whistled. Plaice could be found by treading barefoot for them on the flats around Samson and

Bryher. Fishermen – or more often fisherwomen – walked over the sand feeling for fish with their toes. When they found one the trick was to tread down hard and grab its head before it swam off.

Scillonian plants provided the materials for elixirs and simples. The root of the yellow horned poppy was considered an efficacious remedy for lung complaints. Some tonics were less attractive. There's a revolting recipe which suggests taking two or three eggs and marinating them in lemon juice until the shells have dissolved. Brandy and other ingredients are then added and the resultant gloop bottled for future use.

Scillonians maintained the islands had either feast or famine, never a comfortable sufficiency. Heath describes a harvest supper at which a parson was so diverted by the variety of food on the table that he composed a fantastical list comparing the individual dishes to the different shapes of the islands as seen on a map. Overleaf is the list (with the original eighteenth century spelling preserved).

The Islands compared to a Feast
 1. St *Mary's*, a Scate.
 2. *Tresco*, a Side of Mutton.
 3. *Bryer*, a dry'd Ling.
 4. *Sampson*, Leg of Veal.
 5. *White Island*, near it, a Sole.
 6. *Annett*, a Lobster.
 7. *Agnes*, a Venison Pasty, near it, half a Goose.
 8. *Tean*, a Capon.
 9. St *Helen's*, a Shoulder of Mutton.
10. Bigger *White Island*, a Bacon Ham.
11. St *Martin's*, a Plumb Pudding.
12. *Great Arthur* ⎫
13. *Great Gannick* ⎬ a Brace of Conies.
14. *Great Ganelly* a Breast of Veal.
15, 16, 17, 18, 19. *Scilly*, *Mincarlo*, *Guahall*, *Innisvouls*, *Northwithel*, Roast-Beef and Steaks.
20. *Little Ganilly*, a Plaice.
21. *Ragged Island*, a Conger.
22, 23, 24, 25, 26, 27. *Nornour, Minewithin, Round Island, Little Gannick, Little Arthur, Rat Island*, Pies and Tarts.

The Rocks, and lesser Islands, lying scattered about these, are as Oisters, Cockles, and Shrimps, for Garnish; and the intermixed surrounding Seas as the flowing Tides of Liquor to drown the Care of the Inhabitants.

Fools in new-style hats and coats

On Scilly tourists are a tide that ebbs and flows with the sea. When the *Scillonian* comes in a spring tide of people and luggage floods Hugh Town and all the tills beep and ring.

Pensioners come in search of pretty flower fields and cream teas. They are disconcerted to find themselves up against the elements, forced to ride on open launches where the spray splashes, made to walk over uncultivated heather and among rocks in the sudden rain and queue in pubs in a boozy fug of cigarette smoke. The best sort rise to the occasion and enjoy it. Every morning of their holidays they sally forth undaunted dressed exactly as they would at home. Among this pioneering group, two grey, permed ladies, unmarried or widows it was hard to tell, turned out each day in dresses, brown nylons and slip-on shoes. In concession to the unreliable weather they wore light macs in flesh pink and lime green topped with plastic headscarves. Thus attired as if for a trip to the shops they clambered from boat to boat, shrieking with laughter when waves wet them. From island to island they hauled their stiff bodies and bad legs, holding on enthusiastically to the nice young boatman's hand as he assisted them to alight on slippery quays, and always accompanied by their old-fashioned, fawn leather handbags, receptacles for spectacles, pills, tissues, powder, a mirror, purses, postcard lists, safety pins and peppermints.

The ladies had all the accoutrements of the elderly but they somehow seemed younger than the many "active retireds" who populate the launches and guesthouses. In search of comfort the actives over-react and muffle themselves in elaborate anoraks and expensive outdoor gear designed for rock climbers or fishermen. One couple – a man and a woman, late sixties – threatened by rain on the twenty-minute trip from Bryher to St Mary's, produced a mountaineer's survival bag to protect knees already sheathed in waterproof Gore-Tex.

You see a lot of active retireds on the launches and quays but never out on the downs or headlands they are so well-equipped to visit. The boat arrives at an island and they disembark and disappear, soaked-up by the cafes and tearooms. Boots designed to negotiate the roughest terrain carry them safely a hundred yards up the slipway, and waterproofs tested on Everest or the North Sea keep off the spits and spots of rain encountered on the way. When the boats are due for the return trip they reappear like zombies from the folds in the dunes, mummified in crackling nylon leisurewear. The couples are bloodless and dumb, they seldom say "thank you" or speak to each other. The men carry the money zipped tight in anti-thief pouches. Needless to say their caution is unnecessary, there is hardly any crime on Scilly and most of it is the boisterous after-hours larking which happens long after the timid are asleep.

Setting out for a day's island hopping c1919. Modern visitors pile into the inter-island launches at this same spot on St Mary's Quay.
Picture by Alexander Gibson.

What do these people want with the Islands?

That was the question posed in 1926 by a young Christopher Isherwood in his uncompleted fragment *Vision of Scilly*. Isherwood embarked on the boat journey from Penzance, faltered and left his characters all at sea off Land's End. The question hangs unanswered at the end of the piece.

Isherwood took it up again two years later in his first novel, *All The Conspirators*, which opens on Scilly. The main characters are two young men who stride around striking attitudes and smoking pipes, indeed the first chapters are as much about pipe smoking as the islands. There is an air of furtive depravity personified by a retired colonel who hides among the rocks to photograph mating puffins.

Isherwood visited Scilly in 1926 with his friend Edward Upward. They stayed at Tregarthen's Hotel on St Mary's only to be advised discreetly by the hotel waiter that there were plenty of young *ladies* available on the island.

What *do* people want with islands?

They want to escape the mainland. They want to go to a place that is different, unindustrialised, close to nature. They want to be simple and free. They want to live in a little kingdom and pretend that they own it. They want a holiday, and when they come home they want to take a bit of it back to remind them. But what to take? Photographs and postcards are fine for arousing envy in others who haven't seen the islands. Trouble is, pictures don't look the same as the real thing and anyway, there's the difficulty of the light that makes photos print washed and pale to mainland eyes. Guide books and videos won't do it, too prescriptive. These people want a talisman, something between a sign and a charm.

There are plenty of souvenirs on sale. You can buy bulbs, T-shirts or canvas bags. There is the usual range of artsy-crafty trash; pots which leak, handmade paper too lumpy to write on, sentimental watercolours, stained-glass roundels of dolphins.

Then there are the natural mementoes, free on the beaches to those who bother to collect them. Chalky cockle shells and rhinous yellow toenail limpets will rattle in drawers for years until chucked out. These things are fine but they lack the touch of human history.

In the last three years almost all the beaches have borne a less romantic spoil. In 1997 a container ship went down off St Mary's and dumped tonnes of plastic wrapping overboard. For months the beaches fluttered with opaque shreds and tatters. From a distance the rubbish could have been seaweed; Scilly beaches are often heaped with kelp which looks like wet cellophane. The *Cita's* ripped cargo was a lightweight parody of nature which dried and flipped uncollected about the sand. No one gleaned this wreckage to decorate necklaces, earrings or picture frames, or to foretell the weather. Instead they agonised about the environmental damage, organised a clean-up operation and trusted the sea to eat the last scraps.

Not everything buried at sea stays there and the water around Scilly is full of the dead. In March 2000 Pete Goss's giant catamaran *Team Philips* broke up north of Scilly and had to be towed to safety off St Mary's by the Lifeboat. Fortunately, no lives were lost. Other vessels have been less lucky. At the Hell Bay Hotel on Bryher you can buy genuine pieces of eight recovered by divers from the wreck of a Spanish ship. For £75 you get a coin, a case and a leaflet. Less expensive are the slave manillas, a bargain at only £10 each. Long after Britain abolished slavery, merchants in Birmingham illegally manufactured chunky bangles for use as barter along the coasts of West Africa. In 1843 a whole shipment of manillas, listed as "brass stops", went down with the *Douro* off the Western Rocks. The bracelets sell quite well to hotel guests seeking a curious memento of their stay. They don't buy them to actually wear – too cumbersome and the metal alloy stains pale skin (most of the visitors to Scilly are white).

Opposite page: The Woodcock family bunching flowers at Lower Rocky Hill, St Mary's. The flower industry was still in its infancy when this picture was taken. Narcissi, or lilies as they are called on Scilly, were picked in bud and allowed to open in the sheds to protect them from wind damage. Pug-faced soleil d'ors looking like clusters of mini fried eggs were the favourite. When gathered together their perfume was so overpowering that the workers sometimes fainted. Nowadays women do most of the bunching and the young men who do the picking won't enter the sheds because they are shy of the raucous joking that goes on inside.
Picture by Alexander Gibson, 1890.

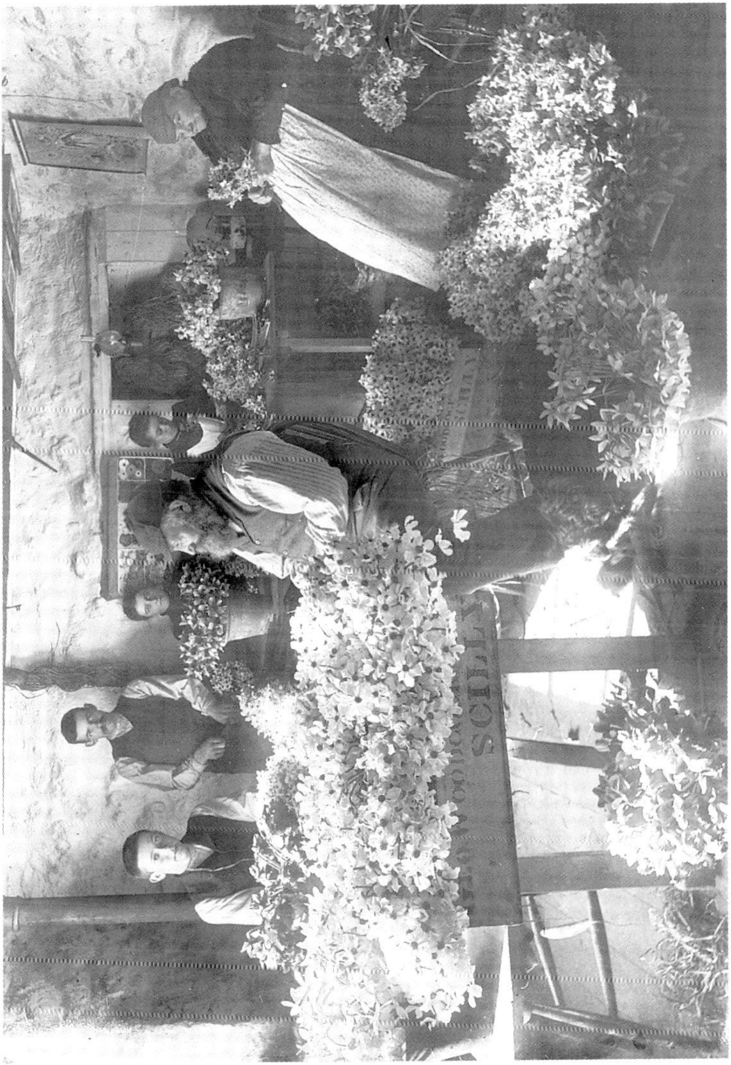

Castaway

In winter Scilly casts itself off from the mainland. From November to March the ferry stops running, the planes and helicopter may be cancelled because of fog or gales. The last flush of visitors departs with the migrating autumn birds. In the frost-free bulb fields scented narcissi begin to bloom and flower-pickers harvest them in curtains of falling rain.

Travel between the islands is unpredictable and the islanders hunker down for the business of survival. The sea-fret goes straight through the walls of the old stone houses and the rooms run wet and green all winter, defeating damp courses, waterproof skins and dehumidifiers. Scilly granite is open-pored and thirsty, it sucks up moisture and is useless for housebuilding. Heath noted that moisture seeping through the stone produced a weird, hard, substance that he said shone like glass, "or hangs pendant from the upper Stones like Isicles." Sensible Scillonians live in ugly modern bungalows such as Harold Wilson's former residence on St Mary's.

Alone on a wide wide sea

The isles' flattened outlines lie low under the sea's horizon. Seen from the deck of a retreating ship they close into one eyelid blink of land. From the air they become a vast bird flopped on the sea with its waterlogged wings outstretched.

Now we have left, Scilly is real island, unreachable, endlessly imagined, longed for,

marooned in memory.

St Agnes lighthouse, taken from a Trinity House receipt of 1680 recording duty paid for maintenance of the light.

Further Reading

BOOKS ABOUT SCILLY:

A Natural and Historical Account of the Islands of Scilly by Robert Heath (1750). Facsimile copies available. Fascinating picture of Scillonian life, the best of the older books.

Observations on the Ancient and Present State of the Islands of Scilly by Dr William Borlase (1756). Borlase's passion for Druidism carries him away.

A Survey of the Ancient and Present State of the Scilly Isles by Rev. John Troutbeck (1794). More druids, plus observations by unpopular Scillonian parson who had to leave the islands after an injudicious involvement in smuggling.

Report detailing the Extreme Miseries of the Off-Islands of Scilly by George Smith (1818). Speaks for itself.

Guide to the Isles of Scilly by J.C and R.W. Tonkin (F. Rodda, 1887). Charming Victorian account with delightful original engravings. A collector's item.

The Scilly Isles by Geoffrey Grigson (Paul Elek, 1948). Still the best guidebook about Scilly, even though Grigson makes the egregious mistake of calling the islands "the Scilly Isles" rather than the Isles of Scilly. A small point, but it infuriates islanders.

The Scilly Isles by C.C. Vyvyan (Robert Hale, 1953). Lady Vyvyan makes an interestingly plaintive attempt to write about Scilly's intangible qualities. Whimsical and very much of its time.

The Isles of Scilly by G. Forrester Matthews (George Ronald, 1960). Useful, worthy, but rather dull account of islands' social, economic and constitutional history enlivened by Gibson photographs.

The Fortunate Islands by R.L. Bowley (Wordens of Cornwall, 1968). Many recent and updated editions available. Recommended short history of the islands.

Augustus Smith of Scilly by Elisabeth Inglis-Jones (Faber & Faber, 1969). Traditional-style biography.

Island Camera by John Arlott, Rex Cowan and Frank Gibson (David and Charles, 1972). The best selection of Gibson Archive photographs, reproduction quality is better in the second edition of 1983.

Islands by John Fowles with photographs by Fay Godwin (Jonathan Cape, 1978). Unfairly neglected work never popular with Scillonians. Fowles speculates learnedly about Scilly, Shakespeare, Homer and Greek myth. Hermetic and excitingly obscure. Persuasive account of the possible origins of the word "Scilly".

A Century of Images by Rex Cowan (Andre Deutsch, 1997). Best account of Gibson family history and lavish selection of archive photographs.

BOOKS TOUCHING ON SCILLY:
All the Conspirators by Christopher Isherwood (Jonathan Cape, 1928). Isherwood's first novel opens on Scilly. In *Lions and Shadows* (Hogarth Press, 1938) Isherwood recounts how he visited Scilly with Chalmers (Edward Upward) at Easter in 1926. The pair discuss Isherwood's new novel, to be called *Seascape with Figures*, and decide it should start on Scilly. The hotel with the discreet waiter has "excellent beer" and Isherwood says the days on Scilly pass in "the excitement of literary composition and the pleasant haze of beer." The fragment, *Vision of Scilly* (available in *The Mortmere Stories* by Isherwood and Upward, Enitharmon Press 1994), was written around the time of the visit as part of the imaginative game the pair played inventing stories and poems about the fantasy village of Mortmere.

Isles of the Island by S.P.B. Mais (Putnam, 1934). Collection of radio talks about British islands including visit to Scilly by the former schoolteacher of the writer J.R. Ackerley.

Islands for Sale by Donald McCormick (Peter Garnett, 1949). Escapist's guide to buying your own island. Includes Scilly and praises the regime on the Cocos (Keeling) Islands. A piece of social history in many respects.

OTHER BOOKS:
Reflections on a Marine Venus by Lawrence Durrell (Faber & Faber, 1953). Like Fowles, Durrell is horribly out of fashion. Neither really deserves to be. *Venus* is far less pretentious and stodgy than Durrell's novels, sharper and sweeter than his poetry.

My Family and Other Animals by Gerald Durrell (Rupert Hart-Davis, 1956). Surely everyone's read this? Even funnier and more evocative than you remember it to be.

Alfred Wallis, Cornish Primitive Painter by Edwin Mullins (Macdonald 1967). Accepts Wallis' belief that his mother came from Scilly. Lively, with condescending appendix by Ben Nicholson. Account of Wallis' life follows Sven Berlin in *Alfred Wallis – Primitive* (Nicholson and Watson, 1949).

Alfred Wallis by Matthew Gale (St Ives Artists series, Tate Gallery Publishing, 1998). Well-researched, basic account of Wallis' life and art.

The Queen of Whale Cay by Kate Summerscale (Fourth Estate, 1997). Much admired biography of Marion Carstairs.

About the author:

Sara Hudston is a writer and freelance journalist. She was born in 1968 and educated at St John's College, Oxford, where she read English and gained a first. Her other published book, *Victorian Theatricals*, is about popular theatre of the nineteenth century. She writes for a range of national newspapers and magazines, lives in West Dorset and is working on a novel.

Islomania was typeset and printed by Bigwood & Staple of Bridgwater in 12 on 14 point Poliphilus, plus Blado italic and Goudy Modern on 130 gsm art paper. Bigwood & Staple is a long-established family firm with a fine printing tradition, founded by Alfred Bigwood in the mid-nineteenth century. In 1883 Frederick Staple was taken on as an apprentice and by 1894 he had taken over the business which has remained in the Staple family ever since. The press occupies an old grain mill on the banks of the River Parrett in Somerset.

About Agre:

Agre Books publishes non-fiction books of literary merit about South West subjects. Based in Dorset, it covers the South West peninsula.

Agre takes its name from the legend of Actaeon and Diana as told in Ovid's *Metamorphoses*. Ovid names Actaeon's hounds and lists their attributes. "Agre the thicket searcher" was the hound with the keenest nose. Agre Books intends to search the thickets of its distinctly rural region to find interesting truths and intriguing stories.

Titles published or forthcoming include:

Bridgwater – The Parrett's Mouth (£9.99). Poems and notes by James Crowden, plus fifty black and white photographs by Pauline Rook taken to mark the 800th anniversary of the town's charter. This unusual book captures the essence of a community and its surrounding countryside.

The Wheal of Hope. Poems and notes by James Crowden, photographs by George Wright. The closure of South Crofty tin mine marked the end of 3,000 years of Cornish history. Wright and Crowden record the last months of the mine and pay tribute to the miners and their heritage.

To find out more about Agre you can write to Agre Books, Groom's Cottage, Nettlecombe, Bridport, Dorset, DT6 3SS, or visit our website at www.agrebooks.co.uk.